# LEATHERBACK SEA TURTLES

## A MIGRATION STORY

amicus LEARNING

by Lisa Amstutz ▪ illustrated by Howard Gray

AMICUS ILLUSTRATED is published by
Amicus Learning, an imprint of Amcius
P.O. Box 227, Mankato, MN 56002
www.amicuspublishing.us



Editor: Alissa Thielges
Series Designer: Kim Pfeffer
Book Designer: Emily Dietz

Library of Congress Cataloging-in-Publication Data
Names: Amstutz, Lisa J., author. | Gray, Howard, illustrator.
Title: Leatherback sea turtles : a migration story / by Lisa Amstutz; illustrated by Howard Gray.
Description: Mankato, MN : Amicus Illustrated, [2026] | Series: Incredible migrations | Includes bibliographical references. | Audience: Ages 5–10 | Audience: Grades 2–3 | Summary: "Follow the incredible journey of a leatherback sea turtle's migration in this narrative nonfiction picture book that will delight animal and nature lovers and support life science education. Includes a migration map, tips to protect animals and habitats, a glossary, and further resources" —Provided by publisher.
Identifiers: LCCN 2024043401 (print) | LCCN 2024043402 (ebook) | ISBN 9781645492856 (library binding) | ISBN 9781681528090 (paperback) | ISBN 9781645493730 (ebook)
Subjects: LCSH: Leatherback turtle—Juvenile literature. | Leatherback turtle—Migration—Juvenile literature.
Classification: LCC QL666.C546 A47 2026 (print) | LCC QL666.C546 (ebook) | DDC 597.92/89—dc23/eng/20250107
LC record available at https://lccn.loc.gov/2024043401
LC ebook record available at https://lccn.loc.gov/2024043402

**About the Author**
Lisa Amstutz is the author of more than 150 children's books. A former outdoor educator, she holds degrees in biology and environmental science. Lisa enjoys learning fun facts about science and sharing them with kids. She lives on a small farm with her family.

**About the Illustrator**
Howard Gray has illustrated a selection of fiction and non-fiction children's books. He has always considered himself an artist, but with a PhD in dolphin genetics, he has a background in zoology. He is now pursuing his dream career in children's illustration from the picturesque city of Durham, UK. Find out more at www.howardgrayillustrations.com.

*Chomp!* A leatherback sea turtle bites a jellyfish. It is her favorite food. She can eat hundreds each day. Her home near California is rich in prey. The turtle must store up energy. She has a long swim ahead. She will migrate across the Pacific Ocean to lay her eggs.

It is fall, and the days are growing shorter. It is time to head for Indonesia. The turtle's journey may cover 10,000 miles (16,000 kilometers).

On and on the turtle swims. Her shell is shaped like a teardrop. It glides through the water. Strong flippers push her along.

The journey is long. Luckily, this turtle is too big for most predators to eat. A leatherback can reach 8 feet long (2.4 meters) and weigh 2,000 pounds (907 kilograms). But a tiger shark or an orca might attack. Steer clear!

The sea turtle swims on. She surfaces every few minutes to breathe. Dives can take her up to 4,000 feet (1,200 meters) deep. She can stay underwater for 85 minutes when resting.

Is that food? Nope! It's a floating plastic bag. But it looks like a jellyfish. Many sea turtles eat these bags by mistake. This can kill them.

It is July now. The turtle is near the beach where she was born in Indonesia. It is almost time to nest. But first she must find a mate. Soon a male comes near.

As the moon rises, the turtle reaches shore. She rides a wave onto the sand.

The turtle scrapes a hole in the sand with her flippers. She lays about 100 eggs. Then she covers them with sand. She drags her heavy body back to the sea. But she stays close to shore.

After 10 days, she makes another nest. She will do this up to seven times. Sometimes, humans and lights scare her away. By September, her job is done. She begins the year-long journey home. She will feed for a few years before returning.

Under the damp sand, tiny turtles grow inside their shells. In Indonesia, poachers steal many eggs. They are sold to eat. Communities help protect the nests. They clean up beaches and stop people from taking the eggs.

After about two months, the eggs start to hatch. Each hatchling is less than 2 inches (5 centimeters) long. Its shell is tough and rubbery, just like its mother's! The hatchling digs its way up through 2 feet (61 cm) of sand.

The hatchling follows the moonlight. Lights from town can confuse her. Hungry predators lurk, too. Some of her siblings do not make it. But the hatchling makes her way to the sea. She finds a large patch of algae. She will live there until she is bigger. Then she will swim to a deeper part of the ocean.

It will be many years until the young turtle returns to the beach where she is born. She will not lay eggs for 15 to 25 years. But one day, if she is lucky . . .

she will make the long trip home.

# LEATHERBACK SEA TURTLE MIGRATION

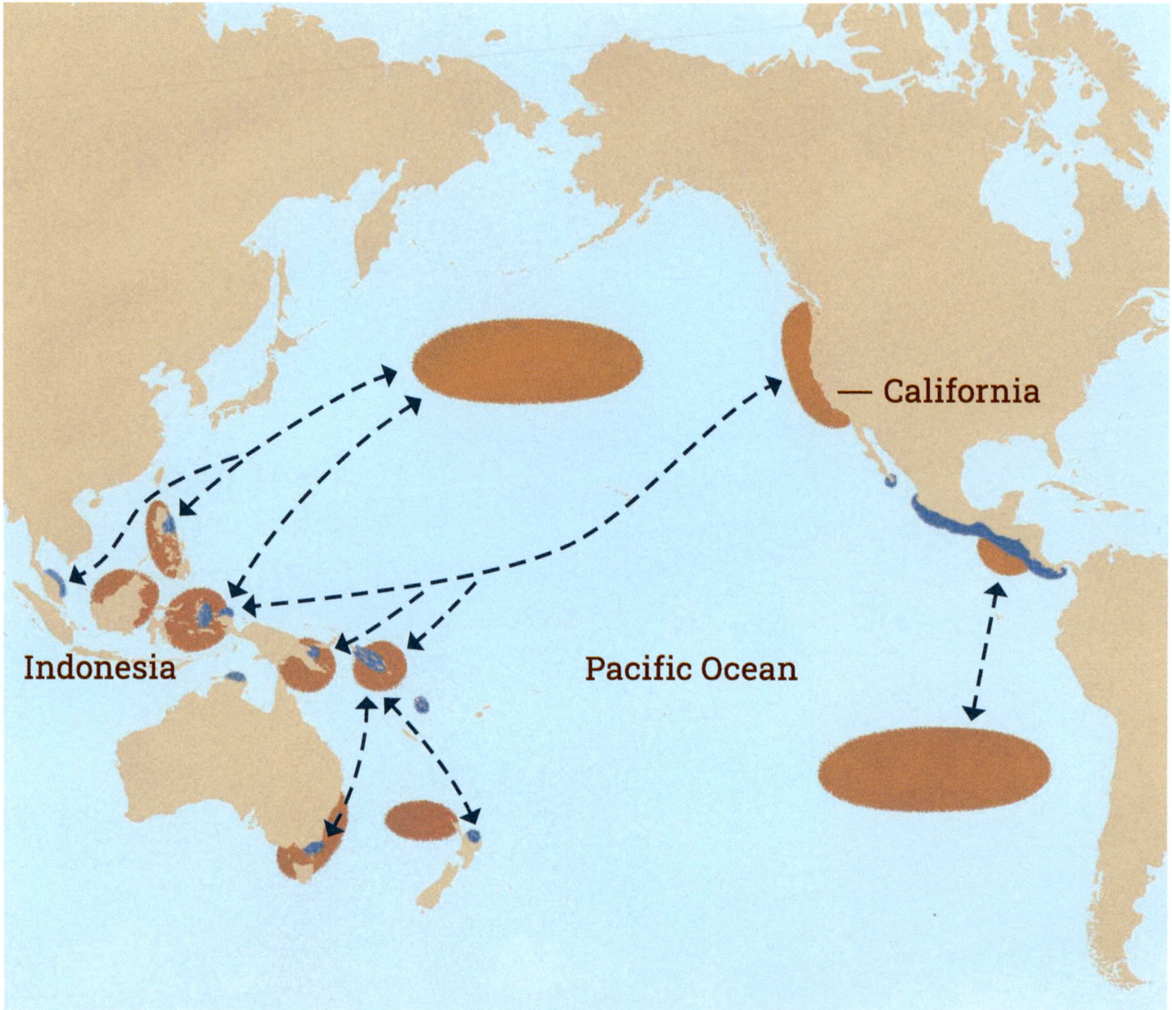


■ Breeding range ← → Migration routes

■ Non-breeding range

## HOW CAN I HELP ?

Humans are sea turtles' biggest danger. Many sea turtles are endangered. You can help by protecting all sea life.

- Help keep beaches clean. Throw away trash.
- Join a beach clean-up.
- Turn off lights near the beach. Lights can confuse nesting turtles and hatchlings.
- Keep plastic out of the ocean. Recycle plastic bags. Better yet, choose reusable ones.
- Help spread the word!

# Glossary

**algae** Simple plants with no leaves or stems that grow in or near water.

**endangered** In danger of dying out.

**hatchling** A young animal that has just come out of its egg.

**mate** The female or male partner of a pair of animals.

**migrate** To move from one place to another to breed or find food.

**poacher** A person who hunts or catches animals illegally.

**predator** An animal that eats other animals.

**prey** An animal hunted or killed for food.

# Read More


Bearzi, Maddalena. ***The Secret Life of a Sea Turtle.*** New York: Roaring Brook Press, 2025.

Donnelly, Rebecca. ***On the Move with Leatherback Sea Turtles.*** Minneapolis: Jump! Inc., 2023.

McDougal, Anna. ***Sea Turtle Migrations.*** New York: Gareth Stevens Publishing, 202


# Websites

**Kids Care About: Sea Turtles**
https://www.timeforkids.com/g56/kids-care-about-sea-turtles/

**National Geographic Kids | Leatherbac Sea Turtle**
https://kids.nationalgeographic.com/animals/reptiles/facts/leatherback-sea-tur